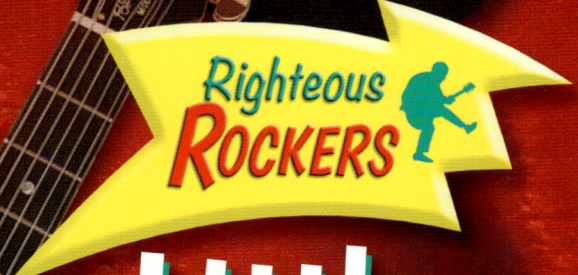

Printing   1   2   3   4   5   6   7   8   9

PUBLISHER'S NOTE
This series, Righteous Rockers, covers racism in United States history and how it affected performers. Some of the events told in this series may be disturbing to young readers.

ABOUT THE AUTHOR: Wayne L. Wilson, novelist and screenwriter, has written numerous biographical and historical books for children and young adults. He received a Master of Arts in Education from UCLA. Wilson is a member of the Writer's Guild of America. He is currently at work on his newest novel.

Library of Congress Cataloging-in-Publication Data
Wilson, Wayne L.
 Little Richard / Written by Wayne L. Wilson
  p. cm.
Includes bibliographic references, glossary, and index.
ISBN 9781624694004
1. Richard, Little, 1932 - — Juvenile literature. 2. Singer — Composer — United States — Biographies — Juvenile literature. 3. African American Musicians — Rock and Roll — Biography — Juvenile literature. I. Series: Righteous Rockers
 ML420.A45 2019
 782.421
[B]
Library of Congress Control Number: 2018943793
ebook ISBN: 9781624694042

Righteous ROCKERS

Bo Diddley
by Nicole K. Orr

Chuck Berry
by Wayne L. Wilson

Fats Domino
by Michael DeMocker

Little Richard
by Wayne L. Wilson

Sam Cooke
by Wayne L. Wilson

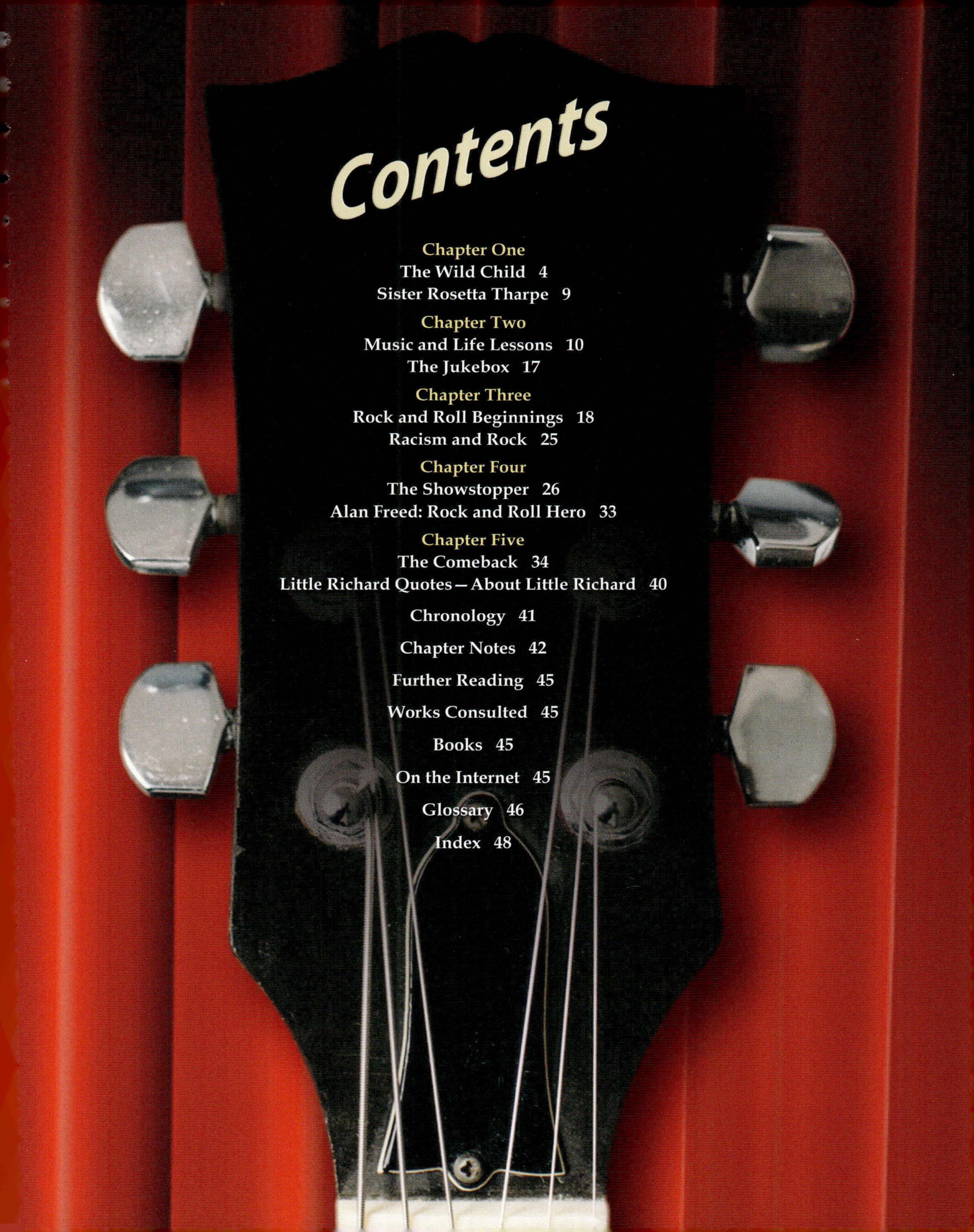

Contents

Chapter One
The Wild Child 4
Sister Rosetta Tharpe 9

Chapter Two
Music and Life Lessons 10
The Jukebox 17

Chapter Three
Rock and Roll Beginnings 18
Racism and Rock 25

Chapter Four
The Showstopper 26
Alan Freed: Rock and Roll Hero 33

Chapter Five
The Comeback 34
Little Richard Quotes—About Little Richard 40

Chronology 41

Chapter Notes 42

Further Reading 45

Works Consulted 45

Books 45

On the Internet 45

Glossary 46

Index 48

THE WILD CHILD

"I was real, real, real shy. It's strange to say it, to see me now."

—*Little Richard*[1]

"Oh no! He's at it again, Leva Mae!" Bud said as he scooted up on the couch, yawning and rubbing his eyes.

"Now, Charles, you know how much that boy loves to sing."

"Yeah, but does he have to sing so loud?"

Leva Mae poked her head out the front door. "Richard?"

The skinny little boy didn't hear her. He bobbed his head and beat the steps to the house and the pots and pans in front of him with two wooden sticks. His eyes were closed and he sang at the top of his lungs.

"RICHARD!"

He opened his eyes and looked at her but kept banging away.

"Your father's trying to take a little afternoon nap before he goes to work. Keep it down."

Little Richard is viewed as one of the most influential figures in the history of popular music and culture. He has often been called the Architect of Rock and Roll.

"Okay, Mama."

He grabbed one of the pans and slowly walked down the dirt street. Once again, he began loudly singing and drumming. Neighbors angrily screamed out their windows:

"Stop it! Ain't nobody wanna hear all that noise!"

"You better get out of here with all that hollering!"

"Boy, will you PLEASE shut up!"

"YOU shut up!" Richard yelled back, as he ran off howling and laughing.

Years later these folks would be amazed to find out that the noisiest kid on the block had become an outrageous performer, making songs that millions paid to hear.

Richard Wayne Penniman was born on December 5, 1932, in a very poor area of Macon, Georgia. His parents were Charles "Bud" Penniman and Leva Mae. His father owned a club and sold bootleg whiskey.[2] Richard was the third child in a family of 12 kids. His parents named him Ricardo Wayne, but a mistake was made on his birth certificate. They decided not to change it since Leva Mae's father's name was Richard.

Richard was born with his right leg shorter than his left. His family nicknamed him "Lil' Richard" because he was so small and skinny. His mother was very protective of him. But that didn't stop others from making fun of him. Some kids thought he was gay and that's why he walked that way. No one, including Richard, knew for sure at that time. Richard recalls, "The kids didn't realize I was crippled. They thought I was trying to twist and walk feminine. But I had to take short steps because I had a little leg. I used to walk with odd strides, like long-

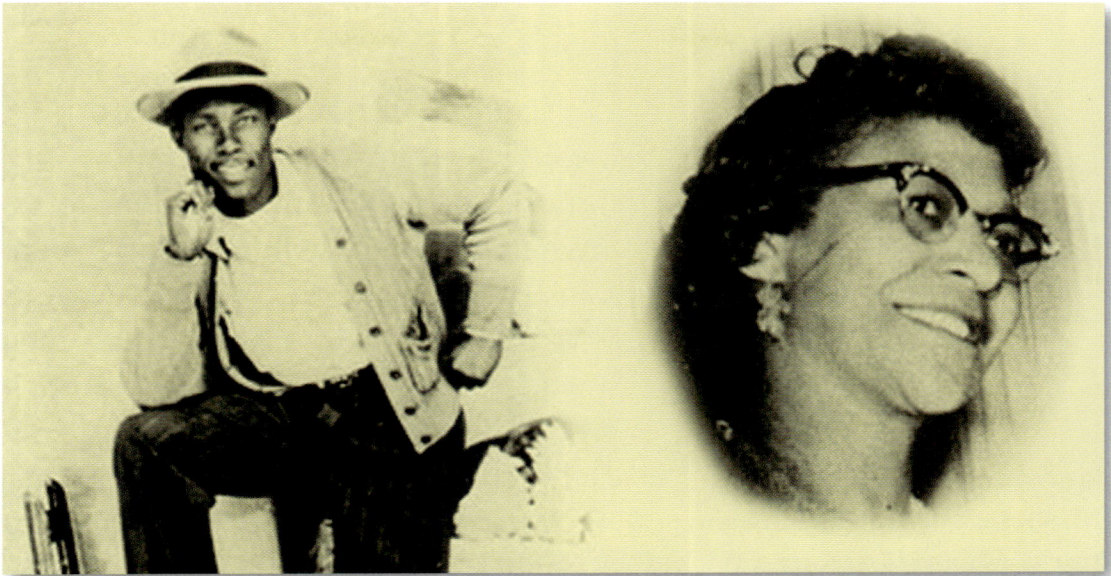

Richard's parents, Charles "Bud" and Leva Mae Penniman, had 12 children. Charles was a church deacon who also sold bootleg liquor and owned a nightclub.

short, long-short. The kids would call me . . . sissy, freak, punk . . . they called me everything."[3]

Richard got into more trouble than any of his brothers and sisters. According to his mother, he was always up to pranks. She punished him a lot because he was constantly breaking the rules. His oldest sibling, Charles, said Richard was always in a playful mood. "He was a showman . . . and it didn't make any difference what he had to do to get attention! Then everybody would get mad—and that's just what he wanted. He'd run away laughing. . . . I used to get mad with him and run after him, but I couldn't catch him!"[4]

However, if Richard got picked on or beat up in a fight, his brother Charles would fight for him. And at home, Richard's oldest sister, Peggy, often took the rap for him. He would do things and then lie about it. Charles and Peggy would be scolded for it. Richard would just stand there and laugh. Charles remarked, "He was good at looking innocent!"[5]

Musicians play in a Macon park. Macon's strong musical heritage influenced such popular musicians as Chuck Berry, James Brown, Otis Redding, and The Allman Brothers Band.

Luckily, Richard took a lot of that wild energy and aimed it toward music. That was an easy thing to do where he lived. Music was everywhere in the streets of this black section of Macon. People sang while they worked, or they sat outside their homes and played instruments. In churches or at town meetings, townspeople joined in singing traditional songs.

Richard remembers women sitting outside washing clothes and scrubbing away on their rubboards. Others might be sweeping the yard. Someone would start up a song and they'd all join in singing. He'd walk up and down the dirt roads in his neighborhood and see people playing guitars. They often sang spirituals and songs about freedom. Richard explains: "See, there was so much poverty, so much prejudice in those days. I imagine people had to sing to feel their connection with God. To sing their trials away, sing their problems away, to make their burdens easier and the load lighter. . . . That's where it started."[6]

Richard was exposed to a lot of gospel music and hymns in church. He learned to play piano and sing with his family in a group called the Penniman Singers. They competed against other groups in gospel music contests. Richard had a naturally loud voice. His brothers and sisters often got upset with him because he'd change the key and sing too loud. Church members called him "War Hawk" because of his hollering and screaming.[7] He was just too loud.

Sister Rosetta Tharpe was born in Arkansas on March 20, 1915, as Rosetta Nubin. Her mother, Katie Bell Nubin, was a singer, a mandolin player, and a preacher for a black evangelical church. Rosetta was a musical prodigy who performed with her mother from the age of four. Rosetta played guitar and was called a "singing and guitar playing miracle."[8]

Mother and daughter moved to Chicago and continued to perform. Rosetta married preacher Thomas Thorpe at the age of 19. The marriage fell apart, but she kept his last name and replaced the "o" with an "a."

Tharpe moved to New York City in 1938. She recorded the first gospel songs for Decca Records: "Rock Me," "That's All," "The Man and I," and "The Lonesome Road." It was the first time a gospel singer received high praise from a non-religious audience. Tharpe performed at Carnegie Hall and shocked and wowed the audience as she played guitar beside blues and jazz musicians. She often played at Harlem's famous Cotton Club with Cab Calloway and Duke Ellington. "That's All" was Tharpe's first recording on the electric guitar, and she became known for this style.

A superstar by the 1940s, Tharpe's crowd-pleasing albums produced many hits. Her powerful voice and remarkable guitar style bridged the gap between gospel and secular music. She inspired such rockers as Chuck Berry, Little Richard, and Elvis Presley.

Rosetta died in 1973. Her headstone reads: "She would sing until you cried and then she would sing until you danced for joy. She helped to keep the church alive and the saints rejoicing."[9]

Rosetta Tharpe, with her mastery of the Gibson guitar, equalled the best of her fellow guitarists.

"People called rock and roll 'African music.' They called it 'voodoo music.' They said that it would drive the kids insane. They said that it was just a flash in the pan."

—Little Richard[1]

Like most of his family, Richard was very involved with the church. He wanted to be a preacher. Richard's grandfather was a minister and his uncle was the pastor of a Baptist church in Philadelphia. His cousin was a minister in a Pentecostal church.

Of all the churches Richard visited, his favorite was the Pentecostal Church because of its joyous and uplifting music. He loved doing the holy dance with the congregation. He tried to imitate people "talking in tongues" (a Pentecostal celebration) even though he didn't know what they were saying. He wanted to be a gospel performer like Brother Joe May, "the singing evangelist" whom people called the "Thunderbolt of the West."[2]

Richard Penniman attended Hudson High School in Macon. He was a poor student, except when it

Richard Wayne Penniman was raised in a Seventh Day Adventist church but he preferred the Pentecostal Church his relatives attended. He loved the congregation's music, holy dancing, healing, and ceremonies. It is also where he learned how to play the piano.

Richard attended Ballard-Hudson High School, which was the only school for African-American students for grades 9-12. Another famous alumni was soul singer Otis Redding (above). His hits include "Respect" and "(Sittin' On) The Dock Of The Bay."

came to music. He quickly learned how to play the alto saxophone. Leva Mae remembered: "That sax has given me many a headache. He used to come home from school and blow it and you could hear it three or four streets away. He could blow it, though . . . he was in the [marching] band right away—and he was real good! I was glad because before that he used to beat on the steps of the house and on tin cans and pots and pans or whatever and sing. He could really sing. But, oh my, the noise."[3]

During this period, Richard was like a sponge, soaking up all types of music—gospel, blues, folk, and country from the music world in the African American community. He continued to sing wherever he could. In high school, he got a part-time job at Macon City Auditorium, working for a show promoter named Clint Brantley. He sold Coca Cola and got a dime for every

Cab Calloway was a very popular jazz singer and bandleader. People loved his unique scat singing and energetic dancing.

bucket of bottles he sold. The best part of the job was getting to see performers like Cab Calloway, Hot Lips Page, Cootie Williams, Lucky Millinder, Marion Williams, and his favorite singer, Sister Rosetta Tharpe.

Once, when he hung around the theater and they set up equipment for the upcoming show, Tharpe walked in. Richard launched into singing two of her hit gospel songs. Impressed, she invited him to sing onstage during her concert. Richard fondly recalls the moment: "Everybody applauded and cheered and it was the best thing that had ever happened to me! Sister Rosetta gave me a handful of money after the show, about thirty-five or forty dollars, and I'd never had so much money in my life before!"[4]

By the time he became a teenager, Richard accepted the fact that he was gay and didn't hide his friendship with other gay people. His homosexuality upset his father, who criticized him harshly. He didn't like the way Richard walked, talked, or dressed, or the type of people he hung out with. Bud told Richard: "My father had seven sons and I wanted seven sons. You've spoiled it, you're only half a son." Richard was hurt. "I couldn't help it. That was the way I was."[5]

His father ordered him to move out of the house. He was only 13. Richard was taken in by a white family, Ann and Johnny Johnson, the owners of Macon's Tick Tock Club. He often performed in their club, and there he perfected his talent.

Richard loved to spend time around the traveling shows that came through town, and sometimes he'd get to perform with them. Then he left Macon to join Dr. Hudson's Medicine Show. Doc Hudson was a snake oil salesman. "Snake oil" describes any mixture that is supposed to cure all kinds of ailments, such as

A re-creation of a snake oil salesman. These showmen were often called frauds, hucksters, quacks, and charlatans because they deceived people by selling fake medicine (snake oil) that were supposed to cure all types of illnesses.

arthritis, rheumatism, and leg or foot pain. Richard helped him sell his fake medicine for two dollars a bottle. Hudson's troupes would hit small towns, and at night they would sleep in a field under a tent. Richard sang the only secular song he knew—Louis Jordan's popular "Caldonia."[6]

His time with Doc Hudson was short. Next, Richard hooked up with Ethyl Wynnes, who owned a club called the Winsetta Patio. She gave him a place to stay and often fed him traditional soul food, such as chitterlings and pigs' feet. He became the lead vocalist for the club's band, the B. Brown Orchestra. The band performed in clubs on the "Chitlin Circuit," which were mainly in the South and areas in the

The Jackson 5 (above) were Chitlin' Circuit performers, as were Sam Cooke, Ray Charles, James Brown, Fats Domino, Etta James, Billie Holiday, B.B. King, Richard Pryor, Redd Foxx, and Aretha Franklin.

A 1957 poster announcing an all-star gathering of rhythm and blues performers. The headliner of this Memorial Auditorium event in Chattanooga, Tennessee, was Little Richard and His Band.

Midwest and North. African Americans were safely welcome to perform in these clubs. The station wagon the band traveled in had signs all over it advertising the name "Little Richard." He would use this name for the rest of his professional career.

During the twentieth century, the jukebox shaped musical culture across the country. Jukeboxes were automated music-playing machines that were operated by dropping coins into a slot. The customer pressed a button for his or her music selection. Jukeboxes were found in bars and hangouts such as pizza joints and ice cream parlors.

The first "music-for-money playing device" was invented by Louis Glass and William Arnold. Their "nickel-in-the-slot phonograph" had four listening tubes. The customer inserted a nickel and then cranked a spring-loaded lever to play the tunes. On November 23, 1889, one of these was installed at Palais Royal Saloon in San Francisco.[7] Over time the look and operation of the machine changed as technology improved.

The Automatic Music Instrument Company created the first multi-selection record player in 1927.[8] This machine sounded great. Customers could hear big sounds from solo singers or huge orchestras for the cost of a nickel. During the 1940s and '50s, the popularity of the jukeboxes boomed.

The colorblind jukebox didn't care about segregation. White customers were exposed to the music of black artists they never saw perform. The jukebox helped sell thousands of records for artists like Chuck Berry, Little Richard, Muddy Waters, Bessie Smith, and others.

People loved the jukebox's design, its animated bubble tubes and being able to see the record-changing mechanism.

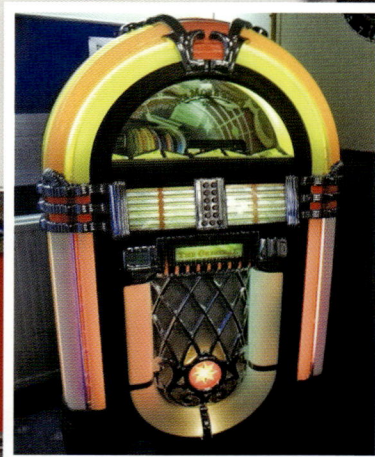

ROCK AND ROLL BEGINNINGS

"Rock and roll offered me a platform to speak what I felt."

—Little Richard[1]

At the age of 15, Richard left the B. Brown Orchestra and joined traveling shows that featured comedy, music, and dancing. One night, one of the girls was out, so Richard took her place. He performed in a red evening gown with high heels under the name Princess Lavonne. He didn't know how to walk in women's shoes, so the band carried him to the microphone and sat him down before the curtains opened. It was common in these types of shows to find men in women's costumes, but his stage act was unsuccessful. Richard quit and headed to Atlanta.

In Atlanta, Richard met entertainer Billy Wright, who strongly influenced his later style. Wright wore loud outfits and matching shoes. He curled his hair up high and used facial makeup. Wright was a gospel blues shouter known in the area as the "Prince of the Blues." He had four Top Ten rhythm and blues

Billy Wright was a popular gospel and blues singer from Atlanta, Georgia. He often wore makeup, styled his hair in a pompadour, and dressed in flamboyant clothes. Wright became a major influence on Little Richard at the beginning of his career.

(R&B) hits. Richard exclaimed, "He was the most fantastic entertainer I had ever seen!"[2]

In 1951, Wright helped eighteen-year-old Richard get his first recording break. He put him in touch with a white disc jockey named Zenas Sears who had a popular black radio program. Sears was able to get Richard a recording contract with RCA. Richard cut four tunes for the company. One was his own composition called "Every Hour."[3]

After the recording session, Richard returned home to Macon.

Although the blues songs Richard recorded did not sell well, they were a hit in his family. When "Every Hour" was played on local radio, everyone in the family got excited, especially his father. Richard said, "My daddy was proud of me for the first time in his life. He made sure that 'Every Hour' was constantly played on the jukebox in his club, the Tip In Inn. . . . He thought I was famous then, but I wasn't."[4]

Eskew Reeder, Jr., later went by the stage name Esquerita. He was an openly gay recording artist who inspired Little Richard and taught him how to play a mean piano.

Another person who influenced Richard in those early years was a teenage musician named Esquerita. (He was born Eskew Reeder Jr.) He gave Richard piano lessons and taught him the pounding piano style he later developed with a vengeance. Richard called Esquerita "one of the greatest pianists. . . . I learned a whole lot about phrasing from him."[5] Richard also modeled his look after Esquerita's, wearing eyeliner, heavy makeup, high teased hair known as a pompadour, and flashy clothing.

Sadly, in 1952, Richard's father was shot dead in a confrontation outside his club. Richard got a regular job to help out his struggling family. He worked as a dishwasher at the Greyhound bus station in Macon. Meanwhile, he continued to play in different bands on the Chitlin Circuit. But he never made enough money with those bands, so he had to keep coming back to the job as a dishwasher.

In 1955, New Orleans R&B singer Lloyd Price suggested Richard send a demo tape to the owner of Specialty Records—Art Rupe. Price had scored the company's first big-selling hit, "Lawdy Miss Clawdy." Rupe found Richard's tape in a brown paper bag with

Lloyd Price's first song, "Lawdy Miss Clawdy" became a major hit on Specialty Records. More hits followed, and he was inducted into the Rock and Roll Hall of Fame in 1998.

grease stains on it. He listened to Little Richard and his band the Upsetters, but the music didn't move him. He tossed the demo into a pile of other tapes.

Months later, he dug out the tape and listened to it again with music producer Robert "Bumps" Blackwell. "The reason we listened to it a second time was that Richard kept calling and bothering us," Rupe recalled.[6] This time they decided they liked his gospel sound and energy. Ray Charles had a new hit, "I Got a Woman." They wanted someone raw and edgy like him. A contract was sent to Richard's old boss Clint Brantley. They set up a recording session for Richard at well-

Blind since the age of seven, the legendary Ray Charles produced such hits as "I Got a Woman," "Hit the Road, Jack," "Georgia on My Mind," and "What'd I Say."

At the age of 18, Matassa open the J&M Recording Studio at the back of his family's shop on Rampart Street in the French Quarter of New Orleans. Matassa became one of the world's top recording engineers.

known recording engineer Cosimo Matassa's studio in New Orleans. Some of the city's top musicians were brought in to make the record.

The session began in September 1955. After two days Bumps was frustrated by how poorly it was going. Richard sang the bluesy songs with little energy. His wild look and the boring songs didn't match.

Bumps decided they needed a lunch break. He took Richard and the musicians to the Dew Drop Inn. When they got there Richard's eyes widened. There on stage was an old upright piano! He leaped onto the bandstand, pounded out a piano introduction, and belted out, "Awop-bop-a-loo-bop-alop-bam-boom—Tutti Frutti!" Richard whooped, hollered, and danced while he played a funky piano accompaniment.

The title for Little Richard's hit song comes from a wild ice cream flavor. Tutti frutti ice cream commonly contains chopped candied fruits, such as cherries, raisins, and pineapple, along with nuts. Sometimes fruit juices are used to add more fruit flavors.

The song was a rowdy tune he used to sing at nightclubs. Bumps loved it! This was the sound he was looking for. He hustled everyone back into the studio. He called a local songwriter, Dorothy LaBostrie, to clean up the lyrics. Drummer Earl Palmer added a backbeat that made the song truly rock. Music scholars credit Palmer as the inventor of rock and roll drumming.[7] Tutti Frutti became a huge hit and rose to number 2 on the R&B charts and number 17 on the Billboard pop charts.

It was the first in a string of monstrous hits for Little Richard.

When rock and roll hit the scene, many groups believed it was destroying society. They blamed the music on African Americans and wanted to ban it. In 1956, Asa Carter, executive secretary of the North Alabama White Citizens' Council, headed a campaign in Alabama to wipe rock and roll off the jukeboxes. He complained the records had dirty lyrics and were immoral. He called rock and roll "the basic, heavy-beat music of Negroes." He also said rock music was a plot by the National Association for the Advancement of Colored People (NAACP) to pull the white man "down to the level of the Negro."[8]

One of the posters handed out by the Council read:

NOTICE! STOP! Help Save the Youth of America. DON'T BUY NEGRO RECORDS! If you don't want to serve Negroes in your place of business, then do not have negro records on your juke box or listen to Negro records on the radio. The screaming, idiotic words, and savage music of these records are undermining the morals of our white youth in America. Call the advertisers of the radio stations that play this type of music and complain to them![9]

Carter later left the White Citizens' Counsel and became a leader in the Ku Klux Klan, an openly racist organization.

Asa Earl Carter was a staunch segregationist.

THE SHOWSTOPPER

"My music made your liver quiver, your bladder spatter, your knees freeze. And your big toe shoot right up in your boot!"

—*Little Richard*[1]

As "Tutti Frutti" climbed up the R&B charts, it was covered by two white artists—Nashville schoolteacher Pat Boone and the wildly popular Elvis Presley. It was common for white record companies to have their artists cover black songs.[2] White radio stations would not play Little Richard's version of "Tutti Frutti." Instead they played Pat Boone's version. Although it lacked the energy, fun, and soul of Richard's song, Boone's record rose to number one and went gold. Boone's success did help the sales of Richard's record, as many whites who had never heard of rock and roll discovered him. His raw and edgy sound became a hit.

In 1956, Specialty Records released Little Richard's next hit single, "Long Tall Sally." On the flip side was "Slippin' and Slidin'." It became a double-sided hit, landing on the national top twenty chart. "Slidin'" was number 1 on R&B and number 17 on the pop

Elvis Presley was a huge fan of black music and admired Little Richard. When he covered "Tutti Frutti," he began receiving hate mail and was pressured not to perform "race music." Richard said in many interviews he was always grateful to people like Elvis for helping such music to reach the top 40 stations.

chart. "Long Tall Sally" became his biggest hit, rocketing to the number 1 spot on the Billboard Rhythm and Blues chart and staying there for eight weeks. It reached number 6 on Billboard's Hot 100. The success was stunning, as it happened without any airplay on white radio stations.

"Long Tall Sally" is number 55 on *Rolling Stone*'s list of the "500 Greatest Songs of All Time." According to Bumps Blackwell, the song was aimed at Pat Boone, knowing he would want to cover it. Blackwell remarks, "We decided to up the tempo on the follow-up and get the lyrics going so fast that Boone wouldn't be able to get his mouth together to do it!"[3] The song became a rock and roll standard, as hundreds of artists covered it.

America wanted more from this super talent known as Little Richard. He gave it to them with his hard rocking rhythms. From 1956 to 1959, backed by high-energy musicians, Little Richard produced sixteen Top 100 hits. Among them were "Jenny, Jenny," "Keep A-Knockin'," "Good Golly, Miss Molly," "Rip It Up," "Lucille," "She's Got It," "Heebie-Jeebies," and "Send Me Some Lovin'."[4]

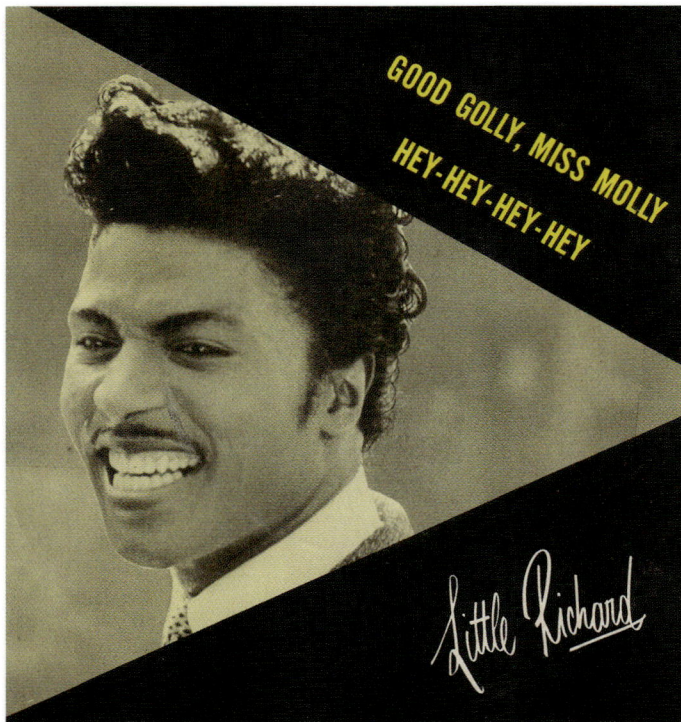

GOOD GOLLY, MISS MOLLY
HEY-HEY-HEY-HEY

Little Richard

Little Richard released this seven-inch 45-rpm vinyl record in 1958.

Richard now had fame, gold records, and enough money to buy a house in Hollywood next door to world champion boxer Joe Louis. He moved his mother and siblings out of their tiny house in Macon to live with him in California. One of his happiest moments was seeing his mother's face when she saw the house. As he described it, "She had never seen black people living in this type of house. It was the kind of house white film stars lived in—big staircase, chandeliers, marble floors, plants, bedrooms upstairs and downstairs, and statues. Really lavish. Until they saw the house and my 1956 gold Fleetwood Cadillac in the garage they never realized what a big hit record really meant. And I didn't either. None of us will ever forget that."[5]

Thirteen-year-old singing sensation Frankie Lymon photographed backstage with Little Richard in 1958. New Yorker Lymon and his group The Teenagers were riding high on their number one "Why Do Fools Fall in Love?"

Richard appeared in three rock and roll movies: *Don't Knock the Rock* (1956), *The Girl Can't Help It* (1956—for which he wrote the title song), and *Mister Rock and Roll* (1957). People loved this hollering black man with the six-inch pompadour who wore brightly colored capes, blousy shirts, mascara and lipstick, and suits studded with sequins. He played the piano like he had a raging fever, often placing his foot over the keyboard or jumping on top of the instrument. His showstopping live performances were incredible!

Little Richard captured America's attention. He was in high demand even in Southern states where Jim Crow ruled. People wanted to see this weirdly cool person who created such a crazy and joyful new sound. Little Richard tore down taboos against blacks appearing in white clubs and dance halls. His shows were filled with mixed-race audiences. In segregated places where blacks were in the balcony and whites on the main floor, the audiences came together and danced.

But during a performance in El Paso, Texas, the police stopped the show. They put Richard in jail because he wore his hair long and danced and shook his body on stage.[6] Still, Richard and his band the Upsetters kept breaking through racial barriers.

Richard said, "The white kids had to hide my records . . .

Fame did not prevent police from harassing and jailing men and women of color. Dr. Martin Luther King Jr. is arrested and pushed against a police desk in 1958 for loitering while trying to attend a hearing for civil rights leader Ralph Abernathy.

During the 1950s there was a huge explosion of school bands forming across the country. They played teenagers' favorite rock and roll songs, like they did at this Valentine's Day dance.

they dare not let their parents know they had them in the house. We decided that my image should be crazy and way-out so that the adults would think I was harmless. I'd appear in one show dressed as the Queen of England and in the next as the pope."[7]

Although the band hated it, Richard made them wear pancake makeup so that they would appear to be gay. He believed they would not be allowed to play in the white clubs if they were seen as a threat to white girls.[8]

Young people were always super excited to go to a Little Richard show. The music made them feel free to do anything they wanted—scream, shout, sing, dance, and jump up and down.

No one wanted to play after a Little Richard performance. He was a master showman and one of the world's greatest musical icons. When he stepped onstage and started playing, people erupted in a dance frenzy within minutes.

During a concert at the Royal Theater in Baltimore, security had to hold people back from jumping off the balcony. The show was stopped twice to remove hysterical fans who were climbing onstage and trying to rip souvenirs from Little Richard's costume.

Richard became a megastar. He achieved the fame he always dreamed of.

That's why it was so shocking when he suddenly quit the music business.

ALAN FREED: ROCK AND ROLL HERO

Super disc jockey and promoter Alan Freed is widely credited for coming up with the phrase "rock and roll." On his radio show in Cleveland in 1951, Freed played R&B music rarely played on mainstream radio. He'd say, "Yeah, daddy, let's rock and roll!"[10] His show was popular among black and white kids.

He left Cleveland in 1954 for New York, where he called his late-night show "Rock and Roll Party." He staged big concerts in New York that drew white and black youth. Parent groups, church leaders, and the press who were anti-rock music started calling him terrible names and accusing him of being a race mixer. But the courageous Freed still put on big dance concerts and countrywide tours. His shows featured artists such as the Drifters, Fats Domino, Joe Turner, the Chantels, Ruth Brown, Chuck Berry, Jerry Lee Lewis, Little Richard, Buddy Holly, and others.

Freed was harassed by citizen groups, the media, and law enforcement, but it didn't stop him from promoting the music he loved. He refused to play R&B songs by cover artists such as Pat Boone when the original music was so much better. He criticized deejays who did so: "They're anti-Negro. If it isn't that, what is it?" he said.[11]

Alan Freed was one of the most popular and influential disc jockeys in music history.

THE COMEBACK

"I am the innovator. I am the originator. I am the emancipator. I am the architect of rock and roll!"
—*Little Richard*[1]

In 1957, Little Richard was in the middle of an Australian tour when he quit rock and roll. Forty thousand Australian teenagers had come to the Sydney outdoor arena to see him. That night Russia launched its first Sputnik spacecraft. The huge bright red ball of fire roared directly over the stadium. Richard took it as a sign. He rose from the piano and announced, "This is it. I am through. I am leaving show business to go back to God."[2]

Richard no longer cared about all the money or fame. He felt something was missing from his life. He even threw an expensive diamond ring into Sydney Harbour. He later found out that the plane he was scheduled to fly in had crashed into the Pacific Ocean. He saw it as another divine sign.[3]

Richard enrolled in the Oakwood Bible College in Huntsville, Alabama, to become a minister. He tried to change his lifestyle. In 1957, he met Ernestine

Still rocking his fans at the age of 75, Little Richard performs at the University of Texas Forty Acres Festival in 2007.

Campbell at a church convention. They married in July 1959. Ernestine claimed they were happily married for a while, but she couldn't handle his international celebrity. "People adored him. Worshipped him . . . I wanted to make him mine and I could not do that."[4] Little Richard admitted to not being a good husband. He said he loved her more like a sister than a wife.[5] They divorced in 1963.

Little Richard disappeared from the public for about five years. He traveled across the country preaching. In 1962 he released the album *Little Richard, King of the Gospel Singers*. Quincy Jones was his musical director. Jones said: "This was Richard Penniman. Little Richard, the rock and roll darling. . . . Gone was the six-inch-high pompadour. Gone the wild mannerisms. . . . Forgotten were 'Tutti Frutti' and 'Long Tall Sally.' Here was a serious young gospel singer. . . . All of us in the studio were deeply moved and impressed."[6]

Soon Richard was doing gospel concerts across the nation. Don Arden, a concert promoter, convinced him to do a tour in Europe, where his records were still selling well.

George Harrison (right), with musician Billy Preston (who sometimes played organ for Little Richard), was a huge Little Richard fan.

Richard arrived in England thinking he was booked as a gospel singer. But his British fans only knew his rock and roll music. The theater was packed and the crowd was excited to see this legendary rock star. Richard walked on stage wearing long religious robes. The great keyboardist Billy Preston (who later gained fame as a singer, too) played organ. Richard sang religious songs.

The audience was confused and disappointed. During intermission,

Arden begged him to sing his rock hits in the second show. They could lose money if Richard continued the tour like this.

Another famous R&B singer, Sam Cooke, performed in the second show and sang all his hit singles, including "Twistin' the Night Away." It was a powerful performance, and the audience gave him a standing ovation. However, Arden was having a fit. If Little Richard followed Cooke's rocking performance with religious songs, the concert could be a disaster. Cooke's manager got him to calm

Singer, songwriter, and businessman Sam Cooke was one of the most popular, beloved, and influential performers in history.

down by assuring him that Little Richard never let anyone outperform him.

As Richard walked onstage for the second show, the lights abruptly shut off, leaving the audience in complete darkness. The only sound in the arena was the low hum of Billy Preston's organ. Then there was a flash from a spotlight, showcasing Little Richard at his grand piano in a bright all-white suit. To the crowd's screams of delight, and with his famous mischievous grin, he immediately began playing "Long Tall Sally." The audience went into a dancing frenzy. They didn't get a chance to catch their breath as Richard played all the hit songs they had been waiting for.[7]

The ovation seemed to last forever. The innovator, the originator, the emancipator, and the architect of rock and roll was back!

Afterward, Little Richard thrilled audiences throughout Europe with his amazing showmanship. Some of his greatest fans were four young

The Beatles meet one of their idols at last, Little Richard.

men from Liverpool called the Beatles. Their singles "Love Me Do" and "Please Please Me" were local hits.

They were delighted to finally meet their idol during his European tour. They often performed "Lucille" and "Long Tall Sally" at shows. Little Richard took the Beatles with him when he toured the clubs in Hamburg, Germany. He talked about his time with John, Paul, George, and Ringo: "They'd come to my dressing room and eat there every night. They hadn't any money so I paid for their food. . . . Paul would come in, sit down, and just look at me. . . . He'd say, 'Oh, Richard! You're my idol.' He wanted to learn my little holler, so we sat at the piano going 'Ooooh! Ooooh!' till he got it."[8] A year later, Richard toured Europe with another as-yet unknown group that also idolized him—the Rolling Stones.

Little Richard returned to America in all his flamboyant glory after enormous success with rock-crazed fans in Europe. Briefly in 1964 he had guitarist Maurice James in his band. James later achieved fame as Jimi Hendrix.

Richard struggled in his comeback. There were minor hits but nothing came close to his earlier dynamic success. Even so, crowds swarmed to see his outrageous live performances at music festivals, casinos, and resorts, many of which were televised.

Richard received high praise for his role in the 1986 movie *Down and Out in Beverly Hills*. It featured his hit single "Great Gosh A-Mighty!"

Little Richard and singer Luther Vandross at Bill Clinton's presidential inauguration in 1993.

That same year he was one of the original inductees into the Rock and Roll Hall of Fame. The honor led to more concerts, television and film appearances, commercials, and soundtrack work. He was on children's shows and earned a gold record for his rock and roll version of "Itsy Bitsy Spider" for a Disney Records album.

In 1990, he returned to Macon for the unveiling of Little Richard Penniman Boulevard. Three years later, he performed at Bill Clinton's presidential inauguration. He received a Lifetime Achievement Award from the National Academy of Recording Arts and Sciences in 1993. In 1994, he was honored by the Rhythm & Blues Foundation with the Pioneer Award. And he was inducted into the NAACP Image Awards Hall of Fame.

Asked about his groundbreaking musical legacy, Little Richard replied: "I think my legacy should be that when I started in show business, there wasn't no such thing as rock and roll. When I started with 'Tutti Frutti,' that's when rock really started rocking."[9]

Little Richard retired from performing in his eighties for health reasons. He was troubled by sciatica and a degenerating hip. At the age of 85, he was designing clothes and reflecting on his monumental impact on rock history.

Many great artists and celebrities have talked about Little Richard's influence in the music world and how much he has meant to them over the years. But nothing is more fun and fascinating to read than Little Richard's comments about himself, taken from several interviews:

To an audience at a Central Park skating rink (1970):

"I am the beautiful Little Richard from way down in Macon, Georgia. . . . Otis Redding is from there, and James Brown's from there, and Wayne Cochran's from there. . . . I was the best lookin' one so I left there first. Prettiest thing in the kitchen, yes sir! I want you to know I am the bronze Liberace! Shut up!"[10]

On being gay (2010):

"I love gay people. I believe I was the founder of gay. I'm the one who started to be so bold tellin' the world! You got to remember my dad put me out of the house because of that; I used to take my mother's curtains and put them on my shoulders . . . called myself the Magnificent One. . . . If you let anybody know you was gay, you was in trouble; so when I came out I didn't care what nobody thought. A lot of people were scared to be with me."[11]

Advice:

"I'd like to give my love to everybody, and let them know that the grass may look greener on the other side, but believe me, it's just as hard to cut."[12]

Little Richard's amazing talent was second to none—so was his humor.

1932   Richard Wayne Penniman is born on December 5.

1945   Richard is kicked out of the house and briefly joins Dr. Hudson's Medicine Show. He then begins performing at the Tick Tock Club.

1947   Sister Rosetta Tharpe hears Richard sing and invites him to sing onstage. At age 15, Richard leaves the B. Brown Orchestra and joins traveling shows featuring comedy, songs, and dancing.

1949   The term *rhythm and blues* is first used by Jerry Wexler of *Billboard* magazine. It replaces the term "*race records*."

1950   The jukebox has become very popular with the dance crowd in bars, diners, and clubs.

1951   Alan Freed begins his program of R&B music on his Cleveland radio station.

1952   Richard's father is killed; Richard takes a job as a dishwasher to support his family.

1955   Little Richard records "Tutti Frutti" with Specialty. It becomes an instant hit.

1956   Little Richard's "Long Tall Sally" is his first number one record on the R&B charts. The North Alabama White Citizens' Council campaigns against rock and roll music because of its black influence. The Council tries to ban it from jukeboxes and radio stations.

1957   Richard shocks his fans by quitting show business. He enrolls in a Bible college to become a minister, then travels the country preaching the gospel.

1959   He releases his first gospel recordings on an album called *God Is Real*.

1962   Richard continues to record gospel albums, including *The King of the Gospel Singers*, accompanied by the Quincy Jones Orchestra. He tours Europe, where his fans convince him to return to secular music.

1963–1970 Little Richard struggles to make a recording comeback, but his live performances draw huge crowds.

1970   He releases "Freedom Blues," a civil rights song.

1986   He is inducted to the Rock and Roll Hall of Fame.

1993   Little Richard receives the Grammy Lifetime Achievement Award.

1994   He receives a Pioneer Award from the Rhythm & Blues Foundation.

2002   He is honored with an NAACP Image Award.

2013   Little Richard announces retirement from show business because of his health.

2017   In a two-hour September interview with Three Angels Broadcasting Network, Little Richard, age 84, discusses his struggles with his sexuality over the years and his vow to pursue his Christian faith in his final years.

2018   Little Richard's hometown in Georgia pledges $90,000 to create a community resource center at Little Richard's boyhood home.

Chapter 1: The Wild Child

1. Darius James, "The Scream: Little Richard Is the Alpha and Omega of American Culture. And He Be Pretty—Very Pretty." *Vibe*. Vol 5, No. 5. Iss: 1070-4701, p. 124.
2. "Little Richard," *The History of Rock*, n.d., https://www.history-of-rock.com/richard.htm
3. Charles White, *The Life And Times of Little Richard: The Authorized Biography* (New York: Omnibus Press, 2003), p. 6.
4. Ibid., pp. 11–12.
5. Ibid., p. 12.
6. White, pp. 15–16.
7. Ibid., p. 16.
8. "Sister Rosetta Tharpe," *Biography*, https://www.biography.com/people/sister-rosetta-tharpe-17172332
9. Roma Panganiban, "The Gospel Singer Who Became the Godmother of Rock and Roll," *Mental Floss*, March 3, 2016 http://mentalfloss.com/article/75377/gospel-singer-who-became-godmother-rock-n-roll

Chapter 2: Music and Life Lessons

1. "100 Greatest Artists: Little Richard," *Rolling Stone*, December 2, 2010. https://www.rollingstone.com/music/lists/100-greatest-artists-of-all-time-19691231/little-richard-20110420
2. Charles White, *The Life and Times of Little Richard: The Authorized Biography* (New York: Omnibus Press, 2003), pp. 16–17.
3. Ibid., p. 18.
4. Ibid, p. 17.
5. Ibid., p. 21.
6. Preston Lauterbach, *The Chitlin Circuit and the Road to Rock and Roll* (New York: W.W. Norton & Company, 2011), p. 215.
7. Casebeer, "Today in History—The First Jukebox Made Its Musical Debut," *American Blues Scene*, November 23, 2013, https://www.americanbluesscene.com/2013/11/today-in-history-the-first-jukebox-made-its-debut/

8. "The History of the Jukebox," *The History of Rock*, n.d., https://www.history-of-rock.com/history_of_the_jukebox.htm

Chapter 3: Rock and Roll Beginnings

1. Jim Booth, "Little Richard: The Ecstasy of Cognitive Dissonance..." *The New Southern Gentleman*, August 27, 2017, https://newsoutherngentleman.wordpress.com/2017/08/27/little-richard-the-ecstasy-of-cognitive-dissonance/
2. Charles White, *The Life and Times of Little Richard: The Authorized Biography* (New York: Omnibus Press, 2003), p. 26.
3. Preston Lauterbach, *The Chitlin Circuit and the Road to Rock and Roll* (New York: W.W. Norton & Company, 2011), p. 216.
4. White, p. 28.
5. Lauterbach, pp. 216–217.
6. Ed Ward, *The History of Rock and Roll: Volume One* 1920–1963 (New York: Flatiron Books, 2016), p. 98.
7. Ibid. p. 108.
8. Linda Martin and Kerry Segrave, *Anti-Rock: The Opposition to Rock and Roll* (Hamden, CT: First Da Capo Press, 1993), p. 41.
9. Michael Rose, "Elvis: The King of Rock and Roll Turns 75, " *Huffpost*, March 18, 2010. http://www.huffingtonpost.com/michael-rose/elvis-the-king-of-rock-n_b_414211.html

Chapter 4: The Showstopper

1. Charles White, *The Life and Times of Little Richard: The Authorized Biography* (New York: Omnibus Press, 2003), p. 125.
2. Ibid., p. 60.
3. Tom Aswell, *Louisiana Rocks! The True Genesis of Rock and Roll* (Gretna, LA: Pelican Publishing Company, 2010), pp. 77–78.
4. 500 Greatest Songs of All Time," *Rolling Stone*, n.d., http://www.rollingstone.com/music/lists/the-500-greatest-songs-of-all-time-20110407/little-richard-long-tall-sally-20110525
5. White, p. 64.
6. Ibid., p. 68.

7. Ibid., p. 66.
8. Aswell, p. 78.
9. Jack Doyle, "Moondog Alan Freed: 1951–1965," *PopHistoryDig.com*, February 28, 2014. http://www.pophistorydig.com/topics/alan-freed-1951-1965/
10. White, pp. 83–84.

Chapter 5: The Comeback

1. Stephen Thomas Erlewine and Keith Harris, "Little Richard: 20 Essential Songs," *Rolling Stone*, n.d., https://www.rollingstone.com/music/pictures/little-richard-20-essential-songs-20160714.
2. Charles White, *The Life and Times of Little Richard: The Authorized Biography* (New York: Omnibus Press, 2003), pp. 91–92.
3. Ibid., p. 92.
4. Ibid., p. 98.
5. Ibid., p. 105.
6. Ibid., p. 103.
7. Ibid., p. 112.
8. Ibid., p. 115–116.
9. Jeff Giles, "Little Richard Announces Retirement: I Am Done," *Ultimate Classic Rock*, n.d., http://ultimateclassicrock.com/little-richard-retiring/
10. David Dalton, "Little Richard: Child of God," *Rolling Stone*, May 28, 1970, https://www.rollingstone.com/music/features/little-richard-child-of-god-19700528
11. John Waters, "When John Waters Met Little Richard," *The Guardian*, November 28, 2010, https://www.theguardian.com/music/2010/nov/28/john-waters-met-little-richard
12. Lepota L. Cosmo, *Rock Love Quote—5000 Quotations on Rock n' Roll of Legends, Bands, Producers, Instrumentalists, Writers and Leading Vocals* (Morrisville, NC: Lulu Press, Inc., 2016), n.p.

Works Consulted

Aswell, Tom. *Louisiana Rocks! The True Genesis of Rock and Roll*. Gretna, LA: Pelican Publishing Company, 2010.

Broven, John. *Record Makers and Breakers: Voices of the Independent Rock and Roll Pioneers*. Chicago: University of Illinois Press, 2009.

Cosmo, Lepota L. *Rock Love Quote—5000 Quotations on Rock n' Roll of Legends, Bands, Producers, Instrumentalists, Writers and Leading Vocals*. Morrisville, NC: Lulu Press, Inc., 2016.

Lauterbach, Preston. *The Chitlin Circuit and the Road to Rock and Roll*. New York: W.W. Norton & Company, 2011.

Martin, Linda, and Kerry Segrave. *Anti-Rock: The Opposition to Rock and Roll*. Hamden, CT: First Da Capo Press, 1993.

Milzoff, Rebecca. "Little Richard on Chuck Berry's Death: 'I Lost One of My Best Friends in Music," *Billboard*, March 21, 2017, http://www.billboard.com/articles/columns/rock/7735494/little-richard-chuck-berry-death-tribute

Ward, Ed. *The History of Rock and Roll: Volume One 1920–1963*. New York: Flatiron Books, 2016.

Waters, John. "When John Waters Met Little Richard," *The Guardian*, November 28, 2010, https://www.theguardian.com/music/2010/nov/28/john-waters-met-little-richard

White, Charles. *The Life and Times of Little Richard: The Authorized Biography*. New York: Omnibus Press, 2003.

Books

Kirby, David. *Little Richard: The Birth of Rock and Roll*. New York: Bloomsbury Academic, 2011.

Reich, Susanna. *Fab Four Friends: The Boys Who Became the Beatles*. New York: Henry Holt and Company, 2015.

Robertson, Robbie. *Legends, Icons & Rebels: Music That Changed the World*. Toronto: Tundra Books, 2013.

Schwartz, Jeffrey. *The Rock & Roll Alphabet*. Los Angeles: Mojo Hand LLC, 2011.

On the Internet

Erlewine, Stephen Thomas, and Keith Harris. "Little Richard: 20 Essential Songs," *Rolling Stone*, http://www.rollingstone.com/music/pictures/little-richard-20-essential-songs-20160714

"Little Richard," *This Day in Music* http://www.thisdayinmusic.com/pages/little_richard

"Tutti Frutti," *Song Facts* http://www.songfacts.com/detail.php?id=1843

amplify (AM-plih-fy)—To make something louder (such as an electric guitar or keyboard).

bootleg—Alcohol drinks, made and sold illegally.

circuit (SIR-kit)—A route through towns and cities, along which performances or other events are held.

confrontation (kon-frun-TAY-shun)—An argument, fight, or challenge.

cover (KUH-ver)—To perform a song that was first performed by someone else.

Creole (KREE-ohl)—A language spoken in the southern United States that is based on French and mixed with words from African languages.

effeminate (eh-FEH-mih-nit)—Showing qualities and mannerisms that are more typical of females than males.

evangelical (ee-van-JEL-ih-kul)—Preaching the words of the Bible and believing that Jesus Christ will save you from sin or hell.

flamboyant (flam-BOY-unt)—Showy; attracting attention by being very lively, colorful, and stylish.

immoral (ih-MOR-ul)—Judged as not good or right; bad.

inspire (in-SPYR)—To cause someone to want to do or create something.

GLOSSARY

Jim Crow—An era in which unfair treatment and laws were passed, mainly in the South, that did not give blacks the same rights as whites.

legendary (LEH-jen-dayr-ee)—Having a history of doing things very well; well-known.

lyrics (LEER-iks)—The words of a song.

phonograph (FOH-noh-graf)—An early record player.

prodigy (PRAH-dih-jee)—A child who is unusually talented.

secular (SEH-kyoo-lur)—Not spiritual or religious.

sequins (SEE-kwins)—Small shiny metal or plastic pieces sewn into clothes for decoration.

taboo (tab-OO)—Not acceptable; forbidden.

technique (tek-NEEK)—A way of doing something that requires special knowledge or skill.

Arden, Don   36–37

B. Brown Orchestra   15, 18

Beatles   38

Berry, Chuck   8, 9, 17, 33

Blackwell, Robert "Bumps"   22, 24–28

Boone, Pat   26, 28, 33

Brantley, Clint   13

Brown, James   8, 15, 40

Brown, Ruth   33

Calloway, Cab   9, 13

Campbell, Ernestine   34, 36

Carter, Asa   25, 26

Chantels   33

Charles, Ray   15, 22

Chitlin' Circuit   15, 21

Clinton, Bill   39

Cooke, Sam   15, 37

Doc Hudson   14

Domino, Fats   15, 33

Drifters   33

Ellington, Duke   9

Franklin, Aretha   15

Freed, Alan   33

Foxx, Redd   15

Gaye, Marvin   15

Harrison, George   36, 38

Hendrix, Jimi   38

Holiday, Billie   15

Holly, Buddy   33

Hudson High School   10, 12

Jackson 5   15

James, Etta   15

Johnson, Ann   14

Johnson, Johnny   14

Jones, Quincy   36

Jordan, Louie   15

Jukeboxes   17

King, B.B.   15

King, Martin Luther, Jr.   30

LaBostrie, Dorothy   24

Lewis, Jerry Lee   33

Lifetime Achievement Award   39

Louis, Joe   29

Lymon, Frankie   29

Matassa, Cosimo   23

May, Brother Joe   10

Millinder, Lucky   13

NAACP Image Award   39

Page, Hot Lips   13

Palmer, Earl   24

Penniman, Charles "Bud" (father)   5–7, 14, 20–21

Penniman, Charles, Jr. (brother)   7

Penniman, Leva Mae (mother)   4–7, 12, 29

Penniman, Peggy (sister)   7

Penniman, Richard (Little Richard)
   arrest   30
   awards   39
   birth   6
   education   10, 12, 34
   hits   28, 39
   marriage   36
   as minister   34, 36–37
   movies   29, 39
   religion   8, 10, 11, 12

Pentecostal Church   10, 11

Presley, Elvis   9, 26, 27

Preston, Billy   36–37

Price, Lloyd   21

Pryor, Richard   15

Reeder, Eskew, Jr ("Esquerita")   20, 21

Rock and Roll Hall of Fame   39

Rolling Stones   38

Rupe, Art   21, 26

Sears, Zenas   20

Smith, Bessie   17

Specialty Records   21, 28

Tharpe, Rosetta   9, 13

Tick Tock Club   14

Turner, Joe   33

Tutti Frutti (ice cream)   24

Vandross, Luther   39

Waters, Muddy   17

Williams, Cootie   13

Williams, Marion

Winsetta Patio   15

Wright, Billy   18–20

Wynnes, Ethyl   15